# Common Sense Guide to Health and Safety in Construction

An essential and short guide for ▁▁▁yees who need to know more about health and safety within the construction industry without wanting to spend hours reading dozens of different documents. Whether it's for use alongside a training course or simply to brush up on your knowledge, it's perfect for equipping you with the principles of health and safety in the construction industry.

Friendly and accessible, this *Common Sense Guide* covers all the main aspects of health and safety in manageable chapters to provide you with the knowledge and understanding you need to look after yourself and others working in construction.

- Suitable for the non-health and safety professional
- Includes questions at the end of each module to consolidate your health and safety knowledge
- Certificate offered to those who complete the exam at the end of the book and return to be marked externally.

**Subash Ludhra** is a past president of The Chartered Institution of Occupational Safety and Health (IOSH) and considered to be an expert in the field of Risk Management. Having qualified as an Occupational Hygienist, Subash Ludhra now manages Anntara Management Ltd; an international risk management and loss control consultancy business that operates in the UK and overseas.

## COMMON SENSE GUIDES TO HEALTH AND SAFETY

Common Sense Guide to Health and Safety at Work
978-0-415-83544-2

Common Sense Guide to Fire Safety and Management
978-0-415-83542-8

Common Sense Guide to Environmental Management
978-0-415-83541-1

Common Sense Guide to International Health and Safety
978-0-415-83540-4

Common Sense Guide to Health and Safety in Construction
978-0-415-83545-9

Common Sense Guide to Health and Safety for the Medical Professional
978-0-415-83546-6

# Common Sense Guide to Health and Safety in Construction

Subash Ludhra

Routledge
Taylor & Francis Group

LONDON AND NEW YORK

First published 2015
by Routledge
2 Park Square, Milton Park, Abingdon, Oxon OX14 4RN

and by Routledge
711 Third Avenue, New York, NY 10017

*Routledge is an imprint of the Taylor & Francis Group,
an informa business*

*British Library Cataloguing in Publication Data*
A catalogue record for this book is available from the British Library

*Library of Congress Cataloging-in-Publication Data*
Ludhra, Subash, author.
  Common sense guide to health and safety in construction/
  Subash Ludhra.
    p. ; cm. — (Common sense guides to health and safety)
  Includes bibliographical references and index.
  I. Title. II. Series: Common sense guides to health and safety.
  [DNLM: 1. Construction Industry—legislation & jurisprudence.
  2. Safety Management—standards. 3. Accidents, Occupational—
  legislation & jurisprudence. 4. Accidents, Occupational—prevention
  & control. 5. Occupational Health. 6. Risk Assessment. WA 485]
  HD9715.S73
  363.11'969—dc23                                    2014011132

ISBN: 978-0-415-83545-9 (pbk)
ISBN: 978-1-315-85878-4 (ebk)

Typeset in Sabon
by Keystroke, Station Road, Codsall, Wolverhampton

# Contents

# Contents

# Foreword

As the Chief Executive Officer of a large farming and food production group, I am only too aware of the vital importance of getting health and safety right, and this includes a keen awareness of the challenges for the construction sector.

In our group, even though it is not our core business, at any time we have a number of construction or renovation projects in motion, and we are conscious of the vital importance of giving time and structure to planning, organising and controlling such work.

As with agriculture, too many people in the construction industry are injured or suffer health problems. Usually the unacceptably high levels of injuries and ill health can be reduced by simple precautions. The hazards are not restricted to those working on sites either, as children and other members of the public can also be affected when construction activities have not been adequately controlled.

It is absolutely clear to me that a safe, well-planned and well-organised construction project is also a much more efficient and effective one, viable and sustainable for the long-term future, protecting people and the working environment.

This new book provides a ready-reference for those who need to know more about health and safety in the construction sector. This guide is offered to support basic training without being overly theoretical or specific. It offers an easy reference for learning the basics of health and safety in construction and the guidance offered is simple but well structured.

It therefore provides a further valuable new learning resource to help us to improve and develop our collective understanding of health and safety, and it follows that any improved learning gives us all an even greater chance to take the actions needed today and tomorrow to avoid the devastation that personal injury causes in construction.

John Shropshire – CEO
The Shropshire Group

# Welcome

It is often said that we all know what is right and what is wrong, what we should and should not do. Surely it's just "common sense". Unfortunately common sense is not as common as we would like to believe. The definition of common sense is "the ability to behave in a sensible way and make practical decisions". Most individuals and employers will at some time do things that are not sensible or practical and some will do this regularly.

This guide has been developed to help improve your knowledge of health and safety in the workplace in a light-hearted way. It is designed to further heighten the common-sense element of your existing knowledge.

Although the guide refers to UK legislation/best practice, the principles are applicable internationally. However, you should answer all questions within the guide, based on the UK legislation.

By taking the time to improve your knowledge and learn more about health and safety in the workplace you can:

- avoid having an accident
- help prevent accidents occurring to others
- make your workplace safer
- make your home safer.

Every year, hundreds of people are killed while at work through accidents and thousands more are killed or suffer from longer-term work-related ill health. The cost to industry runs into billions of pounds and as a result everyone suffers.

Most importantly if you are injured as a result of an accident or suffer from work-related ill health, you and your family are likely to suffer.

## DID YOU KNOW?

- On average 40 people are killed each year while working within the construction industry from accidents.
- About 3,700 occupational cancer cases are estimated to arise each year as a result of past exposures in the construction sector.
- There were an estimated 74,000 total cases and 31,000 new cases of work-related ill health.
- The construction industry has the largest burden of occupational cancer amongst the industrial sectors, over 40% of the occupational cancer deaths and cancer registrations were from construction.
- An estimated 1.4 million working days were lost in 2011/12, 818 thousand due to ill health and 584 thousand due to workplace injury. (HSE 'Contains public sector information published by the Health and Safety Executive and licensed under the Open Government Licence v1.0'.)

By working through this light-hearted guide you will learn more about health and safety in the workplace and as a result you will be better placed to recognise the hazards and dangers in any workplace or even your own home 24 hours a day.

Injured man

# How this guide works

## AIMS

The aim of this guide is to provide you with a basic understanding of:

- the need to manage health and safety
- health and safety law and enforcement
- health and safety legislation
- accidents/incidents and what causes them
- hazards in the workplace and
- proactive measures that can be taken to help reduce accidents to ensure that you can work in a manner that is safe for you and your colleagues at all times.

## OBJECTIVES

By the time you finish this guide, you will be able to:

- identify hazards within your place of work
- define hazard and risk
- help your employer improve health and safety standards at your place of work
- assess risks
- understand civil and criminal law relating to health and safety
- lift and move loads more safely
- be able to recognise and participate in a number of your employer's proactive safety measures
- recognise safety signs and understand their meaning
- avoid taking unnecessary risk
- work safely with electricity
- know when your environment is too noisy
- know what a reportable accident is and who to report it to

- have a greater understanding of health and safety and its importance in your everyday work and home life.

# How to complete the guide

Before you start to complete this guide please read the guidance notes below in order to ensure that you get the most out of your training.

## WHAT DO YOU NEED TO COMPLETE THE GUIDE?

You will need:

- a quiet cosy environment that allows you to relax and make notes
- a pen and paper to make notes
- a desire to learn and improve your knowledge of health and safety.

## GUIDANCE ON LEARNING

This guide has been produced to help you learn more about health and safety in your workplace and to reduce the likelihood of you or your colleagues having an accident or suffering from ill health. The guide allows you to complete your studies at your own pace with the support of your manager or supervisor. However, we recommend that you complete the guide within four weeks.

Throughout the guide there are simple questions designed to help you test your subject knowledge and learn.

If you cannot answer the questions, please read the relevant topic again to refresh your memory. If you are still in doubt, please speak to your line manager who will be able to assist you.

When you have completed the guide and the exercises, you may wish to complete the examination (30-question multiple-choice exam paper) on pages 119–129. On achieving the required pass mark (75%), you will receive a certificate to confirm that you have completed the course and passed the associated examination.

## HOW TO USE THE GUIDE

The guide is divided into modules. We recommend that you complete one module at a time in full, starting with module one, progressing sequentially through to the last module. You do not have to complete the guide in one sitting. A lot of information is provided and you may learn more effectively by tackling the modules in bite-sized chunks.

The modules are designed to take you through a specific learning pattern to help you learn. Each module contains questions to make you think about your own job and workplace. At the end of each module there are a series of questions relating to the module you will have read through to test your knowledge and understanding. The answers can be found on pages 105–110 of the guide. There may be times when you feel you need help and support in completing the guide. Should this be the case please speak to your line manager.

## WHEN DO I GET MY CERTIFICATE?

Once you have completed the guide and the examination paper, the paper will be marked and on achieving the required pass mark a certificate pdf will be emailed to you.

Remember the certificate only confirms that you have completed the guide and passed the associated exam. The real benefit to you will come from your improved knowledge and ability to identify hazards and reduce the risk of having an accident at work or in the home. **Good Luck!**

Winning trophy

# Construction health and safety law and enforcement

# SYDNEY CONSTRUCTION WORKER DIES AFTER FALLING 30 METRES

Welcome to module one. In this module you will learn about the importance of construction health and safety and relevant legislation.

## WHAT IS CONSTRUCTION WORK?

Construction work is defined as:

- Any activity involving alteration, conversion, fitting out, commissioning, renovation, repair, upkeep, redecorating, or other maintenance activity.
- The preparation of an intended structure including site clearance, exploration, investigation and excavation and foundation work.
- The assembly or disassembly of prefabricated elements of a structure, the removal of a structure or part of a structure, or any product or waste resulting from demolition or dismantling of prefabricated elements of a structure.
- The installation, commissioning, maintenance, repair or removal of mechanical, electrical, gas or compressed air, hydraulic telecom, computer or other similar service normally fixed within or to the structure. (HSE – 'Contains public sector information published by the Health and Safety Executive and licensed under the Open Government Licence v1.0'.)

## WHAT IS HEALTH AND SAFETY?

Health and safety is about the measures necessary to control and reduce risks to an acceptable level and to ensure the health and/or safety of anyone who may be affected by the activities of people at work.

"IN REALITY A WORKPLACE CAN NEVER BE 100% RISK FREE."

However, adequate controls and systems need to be in place that are acceptable to the employer, employees and the enforcement authorities.

Any instruction given to you concerning health and safety is for your own well-being and must be adhered to at all times.

Remember successful health and safety management is a two-way commitment between employers and employees and the systems can only be successful when both parties co-operate.

Employers and employees must work together

UK health and safety management revolves primarily around the term REASONABLY PRACTICABLE (i.e. as an employer, did the company do what was reasonably practical to safeguard their employees or others affected by them?).

This is a balancing act of RISK (the probability of an event occurring and the likely consequences if it does occur) against COST (this may be in terms of money, time, effort, latest technology, benefit and so on).

If the RISK relates to slipping in the kitchen and the COST of removing/replacing the slippery surface is very high – then it may be acceptable to minimise the risks in other ways – through warning signs, coatings, the provision of appropriate safety footwear and so on. Generally the greater the risk the less important the cost element becomes.

Risk                                    Cost

Reasonably practicable

## WHY ADDRESS CONSTRUCTION HEALTH AND SAFETY?

### 1. To comply with statutory legislation

You and your employer are required to comply with the requirements of relevant general and specific construction health and safety Acts and Regulations. Examples of Acts and Regulations include:

- The Health and Safety at Work Act
- The Control of Substances Hazardous to Health Regulations
- The Construction (Design and Management) Regulations
- The Work at Height Regulations
- The Personal Protective Equipment at Work Regulations
- The Manual Handling Operations Regulations
- The Provision and Use of Work Equipment Regulations
- The Health and Safety (Consultation with Employees) Regulations
- The Reporting of Injuries, Diseases and Dangerous Occurrences Regulations
- The Health and Safety (First Aid) Regulations.
- The Environmental Protection Act
- The Control of Asbestos Regulations.

This is only a selection of the Acts and Regulations that organisations involved with construction work have to comply with.

### 2. To comply with internal company policies

Responsible employers have internal health and safety policies and procedures. These policies and procedures set out their own internal standards, which in some cases may exceed the requirements of the relevant regulations. You, as an employee, agree to comply with these as part of your terms and conditions of employment.

## 3. Moral obligations

To reduce the number of accidents and incidents and to safeguard you the employee. No employee arrives at work expecting to have an accident. Therefore by increasing your own awareness of health and safety, you can help reduce the likelihood of accidents and incidents occurring and safeguard your well-being.

## 4. It's cost effective

All accidents cost money; for example, staff working overtime, the cost of employing agency staff, loss of production, failure to meet service requirements and extra pressure on the staff left to cover absenteeism. However, the real cost is the damage to the injured employee's well-being and their suffering after the accident. This also affects their family and relations, who are sometimes overlooked after an accident occurs.

## 5. There is a business need to do so

To maintain your employer's good reputation, more and more companies require their suppliers/contractors to have good health and safety systems in place and to be able to demonstrate the promotion of a good, safe working environment.

"A POOR SAFETY REPUTATION COULD COST YOUR EMPLOYER'S BUSINESS DEARLY."

Failing to manage your health and safety can cost you your business

## THE HEALTH AND SAFETY AT WORK ACT 1974

The Health and Safety at Work Act was introduced in 1974 to help protect the large number of employees who, at the time, were not covered by any health and safety legislation. Its aims were:

- to secure the health, safety and welfare of persons at work;
- to protect other people from health and safety risks caused by their work activities – for example, contractors/visitors on site;
- to control the storage and use of explosive, flammable and dangerous substances (this has developed to become the Control of Substances Hazardous to Health Regulations);

- to control atmospheric emissions of certain substances that could prove to be harmful (which has now developed into the Environmental Protection Act).

More importantly it set out clear definitions of responsibilities for employers, employees, the self-employed and suppliers/manufacturers.

## EMPLOYERS' DUTIES

Employers have a duty under the Health and Safety at Work Act to:

- make provision and maintenance of plant and systems of work that are safe and without risk to health;
- ensure safety and absence of risks to health in connection with the use, handling, storage and transport of articles and substances;
- provide information, instruction, training and supervision to ensure the health and safety at work of employees;
- provide you with a place of work, which is in a condition that is safe with respect to access and egress;
- provide and maintain a working environment with adequate welfare facilities such as toilets/ washing facilities;
- ensure that persons not in his employment, that is, visitors/contractors and so on, are not exposed to risks to their health and safety.

Obey the law and your duties

## EMPLOYEES' DUTIES

As well as employers you as an employee have a legal and moral obligation to:

- care for the health and safety of yourself and of other persons who may be affected by what you do or do not do;
- co-operate with your employer by complying with internal and external policies imposed for work activities to ensure your employer is able to fulfil his legal obligations;
- not intentionally or recklessly interfere with or misuse anything provided in the interests of health, safety or welfare. Your employer may have provided safety equipment and procedures to help reduce risk in the workplace and your co-operation is required at all times.

Failing to meet your legal obligations could lead to internal disciplinary action and/or external enforcement action directly against any employee who is found to have broken a health and safety rule that results in injury or property damage.

Tampering with or damaging safety equipment (e.g. machinery guarding, personal protective equipment, fire extinguishers) could lead to dismissal by your employer or prosecution from external enforcement authorities.

> "REMEMBER: WORKING SAFELY MUST BE A CONDITION OF EMPLOYMENT."

## DUTIES OF THE SELF-EMPLOYED

The self-employed have a duty to ensure that their work activities do not endanger themselves or others.

## SUPPLIERS/MANUFACTURERS

Suppliers and manufacturers of articles and substances have a duty to:

- ensure that the product designed or constructed is safe when properly used;
- test or have the product tested to ensure that it is safe;
- provide information and instructions for the user.

## THE CONSTRUCTION (DESIGN AND MANAGEMENT) REGULATIONS

These regulations place duties specifically on clients and other parties (e.g. designers and contractors). They require specific parties to ensure construction activities are properly planned, funded and executed within reasonable timescales using competent contractors and that longer-term, ongoing maintenance and repair are considered during design and construction.

In short they require construction work to be properly thought out and managed.

Some specific construction activities are also subject to prior notification to the health and safety executive.

The regulations also require relevant parties to ensure that suitable welfare arrangement are in place for the duration of the construction work.

## HEALTH AND SAFETY LAW

There are two important areas of law in health and safety.

**Civil Law** Tends to deal with the area of compensation awarded or claimed when incidents occur or something goes wrong.

Magistrates' court

**Criminal Law** Deals with the punishment awarded by the courts when the employer/employee breaks the law.

## LAW ENFORCEMENT

The policing and monitoring of construction health and safety in the UK is primarily carried out and enforced by HSE inspectors.

They have a wide range of powers, which include:

- visiting a place of work at any reasonable time
- carrying out investigations and examination
- dismantling and removing equipment
- taking samples of products
- taking photographs and viewing documents
- taking statements from key personnel
- requiring assistance.

Their primary function is to help to ensure employers are complying with the relevant regulations and safe-guarding their employees and others affected by their business. However, while in their enforcement role they can also:

- issue an **Improvement Notice**. This means something is not safe and they want it made safe, they will stipulate a reasonable time period to allow the employer to comply;
- issue a **Prohibition Notice**. This means that something is so dangerous that the employer/employee must stop doing it immediately. The employer must then make it safe and be able to demonstrate to the enforcement body that it is safe before recommencing the activity;
- **prosecute** employers and employees. This is generally only carried out when there is very strong evidence to suggest that an individual/company deliberately did something they should not have been doing.

## PENALTIES

In a magistrates' court the maximum fine for a health and safety offence is £20,000 (per offence) and/or a **six-month prison sentence**.

In a crown court the maximum fine is **unlimited** and in addition a two year prison sentence can be imposed.

Note: The Health and Safety Executive's inspectors can also charge for their time under the fee for intervention (FFI) scheme where they identify material health and safety breaches.

## HEALTH AND SAFETY ROLES AND RESPONSIBILITIES

As well as the duty imposed on employers and all employees under the Health and Safety at Work Act, within your organisation a number of individuals may have been given specific health and safety duties and responsibilities by virtue of their job functions (e.g. first aider, safety representative, fire marshal). These individuals help to ensure that safety is taken seriously and is properly managed.

**Q**

Do you know who has specific health and safety responsibilities in your workplace? (Write your answer here; if the answer is no then find out.)

# EXERCISE 1

## Self-assessment questions

1. What is the maximum fine you can receive at a magistrates' court?

   a) £10,000
   b) £15,000
   c) £20,000
   d) £30,000

2. What are the two important types of law?

   a) Civil & Matrimonial
   b) Criminal & Peaceful
   c) Civil & Criminal
   d) Peaceful & Matrimonial

3. Can an enforcement officer take equipment away from your workplace?

   a) Yes
   b) No

4. List four reasons for employers to address construction health and safety:

   a) ............................................................................

   b) ............................................................................

   c) ............................................................................

   d) ............................................................................

5. List two parties who have specific responsibilities under the Construction (Design and Management) Regulations.

a) ................................................................................

b) ................................................................................

6. Who enforces construction health and safety law in the UK?

a) Factory inspectors and firemen
b) Policemen and firemen
c) HSE inspectors
d) Firemen and environmental health officers

7. Health and safety law places a duty on:

a) Employers only
b) Employers and employees
c) Employers, employees and the self-employed
d) Nobody

You will find the answers on page 107.

# MODULE TWO

# Construction work issues

# COMPANY DIRECTOR FINED AFTER FAILING TO CONTROL ASBESTOS

Welcome to module two. In this module you will learn more about specific health and safety regulations/requirements that are likely to apply when construction takes place.

## SAFETY SIGNS

You will no doubt have seen signs on display at a construction site and/or other places of work. Signs cost money and give an important message. They are not displayed to make the workplace look pretty.

Signs are produced in four key colours to ease identification. Each colour gives a particular message, which should always be adhered to.

Always know where the signs are, what they mean and most importantly ensure that you comply with their requirements.

First aid

**Dangerous chemicals**

Ear protection must be worn

| COLOUR | MEANING | EXAMPLE |
|--------|---------|---------|
| BLUE | Mandatory (must do) | Wash your hands |
| RED | Prohibition (must not do) | No Smoking/Fire signage |
| YELLOW | Warning of hazards | Caution wet floor |
| GREEN | A safe condition | Fire escape route/First-aid box |

**Q**

Think about some of the signs in your place of work. Can you describe three?

_____

_____

_____

_____

_____

_____

_____

_____

## ELECTRICITY

Electricity is used in almost all construction workplaces and is often referred to as the invisible killer as it can kill without being seen.

Electricity can cause:

- electric shock
- electrical burns
- death (electrocution)
- fires.

The Electricity at Work Regulations were introduced to protect people from dangers associated with electricity and to protect equipment from excesses of electricity.

Employers must ensure that the electrical equipment on site is:

- safe to use
- properly maintained/tested
- easily isolated from the mains supply
- only worked on by competent people
- protected from excesses of current
- protected from adverse weather conditions or other hazards that may be present in a construction environment.

If using electrical equipment you must:

- ensure that the equipment you use is properly maintained
- be competent to use the equipment
- visually inspect it for signs of damage before use
- only use it for what it was designed to be used for.

Remember it is the electrical current (measured in amps) that kills, not the voltage. A very small current passing through your body could kill you.

When using portable electrical appliances, consider using battery-operated equipment or low-voltage equipment (in conjunction with a step-down transformer) along with circuit-breaking devices.

Cables and equipment on construction sites are often subject to greater abuse so must be suitable for their intended environment and may require additional inspections to check condition prior to use.

Remember electrical cables could be hidden below ground, behind walls/partitions, in ceiling voids or be overhead.

In any case of a suspected electric shock to a worker always:

- call for help;
- switch off the power supply if possible and safe to do so, then move the person away from the electrical source using a non-metallic object (do not touch them directly if there is a risk of electrocution to you);
- if the casualty is breathing, place them in the recovery position and call for medical attention;
- if the casualty is not breathing, call for help and attempt resuscitation;
- never touch the casualty with bare hands unless the power supply has been isolated;
- only give first aid if you have been trained to do so.

Also remember that an electrical shock in a construction environment could lead to secondary effects such as a fall from height.

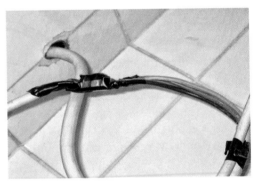

Poor wiring connection

Danger electricty warning sign

**Q**

Do you use any electrical equipment at work or in the home? What precautions do you take prior to and during use?

_____

_____

_____

_____

_____

_____

_____

_____

_____

## OTHER UTILITIES

It is important to be aware of other utilities that may be present on the site. These may include gas or water supplies which could be hidden below ground (pipes or tanks), behind walls/partitions, in ceiling voids or be overhead. In addition there may be compressed air lines, communications cables, oil or fuel supplies used/stored on the premises.

## SUBSTANCES HAZARDOUS TO HEALTH

Substances hazardous to health are any materials, mixtures or compounds used at work or arising from work activities, which have the potential to cause harm to people's health in the form in which they occur in the work activity.

Potential for cement burns to the unprotected hands

Examples include:

- acids used for specialist cleaning
- solvents used to fix floor coverings
- cement in dust or fluid form
- fillers
- glues
- hardeners and other additives.

Less obvious examples include:

- biological diseases from animal/bird faeces
- Weil's disease (from rat's urine)
- contaminated sharps or needles that may have been discarded on the premises
- asbestos used within building materials or sheets.

Asbestos cement sheets dumped on a road

They can be:

- liquids, solids, dusts, powders, gases, or vapours.

They may typically be:

- toxic, corrosive or irritant.

They can cause damage by:

- coming into contact with the skin and eyes
- entering the body through cuts in the skin

- being breathed in
- entering the body through the mouth.

The Control of Substances Hazardous to Health Regulations (COSHH) require your employer to protect you from exposure to hazardous substances by:

- carrying out an adequate assessment of the hazards;
- preventing/controlling your exposure to hazardous substances.

When carrying out the assessment your employer must:

- identify hazardous substances in the workplace
- identify who is at risk
- obtain the relevant material safety data sheets (MSDSs)
- evaluate the risk.

Once the assessments are complete they must:

- eliminate the risk of exposure or introduce appropriate and effective control measures;
- maintain the measures put in place;
- monitor the effectiveness of the control measures;
- record the assessments;
- inform you of all risks and provide suitable instruction, training and supervision.

Control measures that can be applied to reduce risk include:

- elimination
- substitution
- preventing exposure
- limiting exposure time
- providing local or general ventilation
- improving housekeeping

- training
- providing health screening
- providing appropriate personal protective equipment (PPE).

When handling substances hazardous to health always:

- ensure you have been made aware of the hazards and risks;
- store them in correctly labelled containers;
- comply with all control measures deemed necessary;
- treat them with respect.

**Q**

Which substances hazardous to health do you use at work or in the home?

What precautions do you take?

# NOISE

What is noise?

- Noise is generally unwanted sound.
- Noise levels are measured in decibels (dB).
- The human ear's performance generally reduces with age; this process can be hastened by excessive exposure to loud noise over time. Exposure to very loud noises can cause instant damage, even deafness. However, damage is typically caused by exposure to noise over a longer period of time.

The Noise at Work Regulations were introduced to protect people from excesses of noise.

Two key action levels were introduced which are explained below.

## 80dB (A)

If noise levels exceed this (averaged over 8 hours), then your employer is required to:

- carry out a risk assessment (by a competent person);
- make reasonable attempts to reduce noise levels using control measures to below this point. If he cannot, then he is required to inform you, his employees, and offer hearing protection to you if you wish to use it.

## 85dB (A)

If noise levels exceed this (averaged over 8 hours), then hearing loss is likely. If the levels cannot be reduced through engineering controls, then the employer is required to:

- carry out a risk assessment (by a competent person)
- provide suitable hearing protection
- ensure that you wear the protection provided
- mark out hearing protection zones.

The control measures available to your employer to help prevent hearing damage include:

- the use of suitable equipment
- the selection of new quieter equipment
- planned maintenance of equipment
- enclosing the noisy equipment with sound insulating material
- enclosing the workforce in sound insulated booths
- reducing the time employees are exposed to the noise
- providing suitable PPE
- marking out noisy areas as Ear Protection Zones.

| TYPICAL NOISE LEVELS DECIBELS (dB(A)) | |
| --- | --- |
| Jet aircraft (taking off at 25m) | 140 |
| Chainsaw | 120 |
| Heavy traffic | 80 |
| A busy office | 60 |
| A typical library | 20 |

Ear plugs

As a guide, if you are standing two metres away from a colleague and cannot hear them when trying to have a normal conversation, the environment is probably too noisy.

Remember, once hearing is damaged it may never be repaired.

In addition to noise, some work equipment used in the construction sector vibrates and the vibration can be harmful to the body, internal organs or hands (vibration white finger). Examples include riding trucks or using pneumatic drills or other vibrating hand tools.

Employers need to monitor and control the amount of exposure employees receive to ensure compliance with the control of vibration at work regulations.

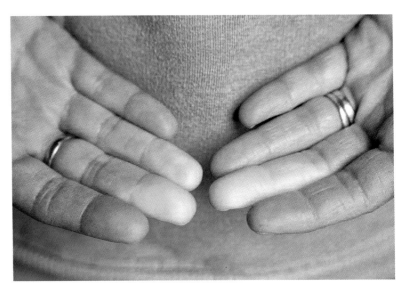

Vibration white finger

What is the noisiest place or machine in your place of work? What control measures does your employer have in place?

## MANUAL HANDLING

Manual handling is the lifting, lowering, pulling, pushing or moving by hand or bodily force an object or load. It is a function you perform each and every day from brushing your teeth to picking up a shopping bag.

Some job functions require more manual handling than others and it's not just the weight of a load that can cause problems. In fact the incorrect manual handling of any load can cause injury.

In the UK millions of working days are lost each year due to pains, strain and other injuries to the back. Manual handling injuries make up about a quarter of all injuries reported to the enforcing authorities.

Injuries can occur from a single event or may develop over a period of time.

The types of injuries suffered from manual handling are not confined to the back and can include cuts, hernias, wrenched shoulders, crushed feet or fingers, fractures and bruises.

The Manual Handling Operations Regulations were introduced to help reduce the number of accidents occurring within the workplace. In simple terms the regulations say that if manual handling can be avoided, then it should be. If it cannot be avoided, then your employer is required to carry out a risk assessment and introduce control measures to minimise the risk of injury to you.

The assessment must take into account:

- the **Task**, that is, what you are actually required to do;
- **Individual capability**, that is, what you are capable of handling;
- the **Load**, that is, what it is you are required to handle;
- the **Environment**, that is, your surroundings.

**Always remember the following procedure:**

1. Examine the load – can the manual handling activity be avoided by redesigning or changing the task or avoided with the use of mechanical aids?
2. Are you physically capable of handling the load or is assistance required?
3. Is protective clothing required to protect your hands, body or feet?

**When lifting a load always:**

1. Face the way you are walking or working – avoid twisting your trunk and overstretching at all times.
2. Position your feet so that they are approximately the width of your hips apart, with one foot slightly in front of the other and flat on the floor to provide a stable, balanced stance. Once the load has been lifted, the weight may be transferred to the front foot.
3. Always ensure you have a firm grip on the load, even if it means using gloves to prevent the load slipping.

4. Your back should be kept straight to maintain it in its natural and strongest position. To get down to the load, the knees and the ankles should be bent and the load raised gradually using the thigh and leg muscles.
5. The head should be kept up and the chin well in, as this helps to keep the spine in its natural upright position.
6. Arms should be kept as close to the body as possible, which helps to retain balance.
7. The body should be used to counterbalance the weight of the load.
8. If it is a team lift, then one person should control the lift so the lift is even and together.

**When carrying a load always ensure:**

1. You can see where you are going. Loads which extend to head height and which obstruct your vision are dangerous to you and to other persons nearby – if this is the case you should either use a mechanical aid (sack barrow or pallet truck) or seek assistance from a colleague.
2. Your route is clear of any obstructions or slippery areas.
3. Particular care is taken when going around corners or negotiating stairs.
4. Sudden movements and twisting of the spine is avoided – face the way you are walking.
5. That if the load cannot be set down in the required position it should be put down temporarily and re-lifted when the position is clear.
6. You do not try to change your grip while carrying – if your grip is slipping put the load down and start again.

**When putting down a load always ensure:**

1. Your back is kept straight and in its natural position. If the load is to be lowered, the thigh and leg muscles must be used.
2. Your head is up and your chin tucked in.
3. You maintain a proper palm grip, but beware of trapping the fingers and hands beneath the load.
4. You use your body weight to counterbalance the load.
5. You keep your arms close to your body. Setting down and stacking should be only as high as it is possible to reach with the elbows tucked into the sides. Do not over-reach.
6. You maintain a firm, balanced stance, with your feet as close as possible to the centre of gravity of the load, and flat on the floor. If the load you are moving is too heavy or bulky always seek assistance, which can be a colleague of similar height and build (one person should call the signals) or mechanical aids.

**Q**

What is the heaviest item of equipment you have to handle and have you been properly trained to handle it?

# PERSONAL PROTECTIVE EQUIPMENT (PPE)

Examples of PPE used on construction sites

PPE for welders

Personal protective equipment is any device or piece of equipment held or worn that can provide protection against one or more hazards.

When used correctly PPE can be used as an effective control measure against hazards. However:

"WEARING PERSONAL PROTECTIVE EQUIPMENT DOES NOT REDUCE THE HAZARD AT ALL. THE HAZARD REMAINS THE SAME EVEN THOUGH THE RISK TO YOU IS REDUCED."

As such it must always be used as a last resort when all other control measures have been exhausted. The Personal Protective Equipment Regulations require your employer to carry out a risk assessment and identify when and where the wearing of PPE will be necessary.

When using PPE (which should be provided at no cost to you), your employer must take into account the following:

- its fit (i.e. is it the right size/shape for you?);
- its suitability against the risk;
- the ergonomics of the PPE (i.e. how well the PPE is designed for the individual user);
- any increase in risk to you or others as a result of using the PPE;
- its ongoing cleaning/maintenance;
- replacement procedures;
- storage facilities;
- adequate training to ensure you understand why it needs to be worn, the benefits it will provide to you, how it is to be maintained and stored and how you obtain replacements.

Questions to ask yourself when in your workplace:

- Is there a need for PPE for the job I am doing?
- Do I need protecting against any hazards?
- If I use PPE is it suitable and does it serve its purpose?
- Is the PPE in good condition and cleaned regularly?
- Does it fit correctly and is it comfortable to work in?
- Is it readily available?
- Can I get replacements?
- Have I been trained on how to use the PPE? (On more complex equipment recorded evidence of training must be documented.)
- Does the PPE cause me any problems while I am working (heat rash or allergies)?
- Where do I store it when I have finished with it?

**EXAMPLES OF PPE**

| Body part protected | Type of PPE available |
|---|---|
| Body | Overalls, coveralls, tabards |
| Feet | Shoes, boots |
| Hands | Gloves, gauntlets, wrist cuffs |
| Ears | Ear plugs, muffs |
| Eyes | Goggles, glasses |
| Face | Visor, hood, shield |
| Head | Bump caps, helmets |
| Skin | Barrier creams |
| Lungs | Breathing apparatus, respirators |
| Legs | Knee pads, leggings |

**Q**

Do you or any of your colleagues wear PPE? If so, what is it worn for and how do you get it replaced?

_____

_____

_____

_____

_____

_____

_____

_____

_____

_____

## WELFARE FACILITIES/YOUR WORKPLACE

Your employer is required to provide you with adequate facilities and ensure the workplace is safe and healthy. This includes controlling access (signing in/out, safety briefings and the issuing of PPE) to the construction site or area and preventing unauthorised access.

### Ventilation

Air must be provided either by natural (e.g. an open window) or mechanical means (e.g. a ducted fan-assisted system). If the construction activity generates dust, then this must too be managed and workers protected accordingly.

Construction site compound

## Temperature

The workplace temperature (within buildings) should be at least 16°C (or at least 13°C where the work involves physical effort). However, if you work outside or in areas where it is unreasonable to maintain the minimum temperature or in areas that require temperatures that have to be kept high or low, then your employer may have to provide you with specialist clothing or make other suitable arrangements. This may include thermal clothing for cold environments or sun protection and shelter when working in direct sunlight.

## Lighting

Your employer is required to ensure that there is suitable and adequate lighting for the tasks that you perform. This may include artificial lighting if working in the hours of darkness or where natural lighting is not available.

## Stairways

These should be kept clear and well lit. Guardrails must be fitted to open stairways.

## Workstations

Your employer must ensure that you have adequate space to perform your job safely.

## Furniture

This must be suitable for the tasks being performed.

## Floors/Walkways

Must be defined, kept clear and in good repair to prevent slips, trips or falls. If they are close to roads or areas of vehicle movement, they may also require additional protection.

## Housekeeping

The workplace must be kept clean and tidy. Equipment should be correctly stored and not left lying around.

## Toilets

An adequate numbers of urinals and water closets (WCs) must be provided at a reasonable distance to where the work is being carried out, taking into account the duration of the work.

## Washbasins

Based on the number of male/female employees present. They must be suitably maintained with soap and cold and warm water available.

## Rest areas

May be required depending on the nature of the construction activity and its duration.

# WORK EQUIPMENT

Work equipment is defined as any machine or hand tool used at work. From this definition it is clear that almost all construction workplaces will have work equipment present. Although work equipment is generally there to make tasks easier, quicker or more efficient, if used incorrectly or without training, the work equipment can be dangerous and cause injury.

Under the Provision and Use of Work Equipment Regulations (PUWER) your employer is required to:

- provide suitable and safe work equipment;
- provide suitable and sufficient training on the correct use (including risks and precautions necessary) of the equipment for all authorised operators of the equipment.

There are a number of specific dangers associated with work equipment. These include:

## Entanglement

Loose clothing or jewellery worn by you could become entangled within the moving parts of the machinery.

## Entrapment

Some machines can entrap parts of your body within their moving parts. Fingers, limbs and even the whole body can be pulled into the machine and crushed.

## Contact

Moving parts of a machine could cause severe injuries to limbs or burns to the skin.

## Ejection

Moving or rotating machines can throw out objects unexpectedly, causing you injury. In addition fluids can be ejected from ruptured pipes or hoses.

## Impact

You could be struck by a moving machine/vehicle or an object being worked on by a machine.

Greater safety when using work equipment can be achieved by:

- buying and only using equipment that has been designed and constructed to remove foreseeable dangers;
- ensuring that equipment is used in a safe place not giving rise to danger;
- guarding dangerous parts of equipment to prevent injuries. Guards are designed and fitted to equipment

to protect you and the equipment. Never tamper with guards, try to override them or operate a machine with the guards removed;

- never wearing loose clothing or jewellery around machinery;
- covering or tying long hair to prevent entanglement;
- ensuring adequate lighting so that everything can be clearly seen;
- ensuring that you wear any personal protective clothing deemed to be necessary;
- switching off and isolating the equipment when not in use and preventing unauthorised use;
- ensuring that the area around the work equipment is clean and tidy to prevent things or people falling;
- reporting all faults or suspected faults to your manager so that they can be investigated and where necessary rectified;
- only using work equipment if you are trained, competent and authorised to do so;
- never distracting other people who are using work equipment.

**When using hand-tools the following rules apply:**

- Visually inspect the tools and cables for signs of damage prior to use and report any defects.
- Always use the correct tool for the job and the work environment. Do not attempt to make the tool fit the job.
- Ensure that the tools are in good condition.
- Use the tools in the correct way.
- Never use tools you have not been trained to use.
- Take out of service any broken or damaged tools identified.
- Work equipment can only ever be as safe as the operator using it.

## DEMOLITION

Construction work may involve the demolition of structures (or parts of structures) and this can be one of the most hazardous activities taking place due to the vast range of potential hazards which can include:

- noise and vibration
- plant/vehicles colliding or overturning
- exposure to utilities (gas, electricity and water)
- dusts, fumes or smoke
- power tools
- falls from height or falling materials/debris
- premature collapses
- exposure to asbestos or other harmful substances
- fires or explosions.
- environmental contamination.

Demolition work should always be properly planned and only carried out by competent persons/contractors. Employees should never attempt to demolish any structure unless they have the necessary competence and have taken appropriate precautions.

## WORK AT HEIGHT

Work at height accounts for a large proportion of deaths or serious injuries within the construction industry.

Under the Work at Height Regulations, employers need to protect persons where there is a risk of falling a distance liable to cause personal injury (there is no minimum height limit). The person could be above ground, on the ground or below ground and still be working at a height under the regulations.

The regulations state that the following hierarchy should be followed

1. Work at height should be avoided (e.g. telescopic rods for window cleaning).
2. If it cannot be avoided, employers need to prevent employees from falling (e.g. the use of guard rails).
3. And lastly protect them if they do fall (e.g. the use of air bags).

Only work at height if you believe that your employer has considered the above hierarchy and taken appropriate precautions. For example:

- properly planned and organised the work;
- taken account of weather conditions;
- trained persons to carry out the work and deemed them to be competent;
- provided a safe place of work;
- provided equipment appropriate to the work and had it inspected;
- considered risks from fragile surfaces and introduced controls;
- considered risks from falling objects and intruded controls.

Remember roofs can be fragile even when they appear to be safe and additional precautions will need to be taken if you are working at height near water (e.g. rivers, lakes, open tanks) due to the added risk of drowning.

## ACCESS EQUIPMENT

### Ladders

It may be necessary to use ladders or stepladders when working at height. These should only be used when the risk is deemed to be low, the duration of the work is likely to be short and existing features of the site make it difficult or impractical to use other types of access equipment.

Whenever using a ladder ensure:

- the user can securely hold the ladder;
- the work is not at an excessive height and the ladder is long enough without the operator having to over-reach;
- the ladder can be secured to a fixed anchor point;
- the work is of short duration;
- the work area allows for the safe use of the ladder;

Ladder access on a scaffold

- the ladder is appropriate for the work being carried out and the environment it's being used in;
- the weather conditions are suitable;
- there is no risk of collision from vehicles or others;
- the ground conditions are suitable;
- the ladder is stable and can achieve an inclination of 1:4;
- the user does not have a fear of heights;
- there are no overhead power lines likely to impede the work.

## Mobile towers

When using mobile towers, employers should ensure:

- the ground is firm level and stable;
- persons erecting the tower are competent;
- appropriate diagonal bracing is in place and where necessary additional stabilisers have been fitted;

Mobile tower with outriggers

- users do not climb the outside of the tower;
- the wheels are locked when being used;
- the tower is not moved while persons are on it;
- towers are not used on inclines or slopes;
- the working platform is fully boarded, has guard rails/kick boards and is not overloaded;
- the tower is regularly inspected especially if exposed to adverse weather or suspected tampering;
- persons on the tower are not overreaching or using ladders, steps or trestles on the platform;
- the tower is appropriate for the work to be carried out.

## Mobile elevated working platforms (MEWP)

Mobile platforms are very versatile and offer users the ability to reach areas not necessarily accessible from ladders or towers. They can operate vertically (e.g. scissor lifts) or three dimensionally (e.g. cherry pickers). However, they do have their own limitations and inherent risks.

When using working platforms, employers should ensure:

- the equipment is only used by trained, competent and authorised persons;
- the users are trained in safe evacuation from the tower in the event of a failure;
- the equipment is properly maintained and inspected by a competent person;
- that the platform is properly protected from vehicles in the vicinity;
- it is only used on level stable ground capable of withstanding the load;
- the working platform is never overloaded;
- the base is not moved while the platform is elevated;
- the tyres are properly inflated and the wheels are immobilised;

An example of a mobile elevated working platform (MEWP)

MEWP

- where outriggers are necessary, they should be fully extended and in the locked position;
- that account has been taken of any overhead power cables.

## TRAFFIC FLOW AND MANAGEMENT

The movement of vehicles and people on construction sites must be properly managed. In addition only appropriate maintained vehicles should be allowed on the site, operated by trained, competent and authorised persons.

Some vehicles will require additional protective devices such as:

- seat belts to prevent drivers being thrown from vehicles;
- falling object protective structures to protect drivers from falling materials;
- roll/tip over protective structures to protect drivers from injury.

Where possible, pedestrians should be separated from vehicle routes using physical barriers and separate entrances for people and vehicles. One-way systems can help reduce the need to reverse vehicles, but where reversing is necessary, then vehicles should be fitted with audible/visual alarms and a banksman should be considered.

Speed limits should be established and enforced and any obstacles/excavations or structures should be protected.

Where there are slopes, banks or gradients that vehicles may have to travel over or near, additional precautions may be necessary to prevent the vehicle overturning.

Walkway created at a construction site

## EXCAVATIONS

When working in or near excavations care must be taken to prevent:

- falls of equipment, vehicles, persons and materials;
- the collapse of the excavation or structures near by;
- damage to buried services;
- dangers associated with flood and hazardous substances.

Precautions necessary may include:

- warning signs
- barriers
- shoring and battering the excavation

- using access ladders
- installing crossing points
- introducing additional lighting
- personal protective equipment
- checks for buried services
- remaining clear of mechanical diggers
- identifying safe routes for equipment, materials and vehicles
- regular inspections
- in filling.

Barriered excavation with some basic support

Excavations being mechanically dug

## CONFINED SPACES

Each year workers are killed carrying out work in confined spaces. A confined space can be any place including a chamber, tank, vat, silo, pit, pipe, sewer, flue, well or other similar space in which by virtue of its enclosed nature there is a foreseeable risk of a specified occurrence.

A specific occurrence may be:

- a fire or explosion;
- loss of consciousness or asphyxiation of any person at work arising from gas, fumes, vapours or lack of oxygen;
- drowning;
- asphyxiation as a result of free-flowing solids;
- loss of consciousness as a result of high temperatures.

When faced with a confined space environment, employers should ensure that:

1. No person works in a confined space where it is reasonably practicable for the work to be carried out without entering the space.
2. Before carrying out the work, point one must be satisfied and an appropriate risk assessment carried out.
3. No person enters, leaves or carries out work in the space without a safe system of work.
4. Emergency procedures have been put into place.

When establishing emergency rescue procedures, the following should be considered:

- What rescue and resuscitation equipment is likely to be necessary?
- Does the local hospital need to be notified?
- How will an alarm be raised and how will the rescue be implemented?
- How will the rescuers be safeguarded?
- Do other third parties need to be safeguarded?

- Will fire-fighting equipment be necessary?
- What equipment needs to be isolated/controlled?
- What first-aid provision will be necessary?
- What emergency contact details and methods of communication will be necessary?
- What local emergency services are available?

**Q**

Can you identify any potential confined spaces that you may have come across at work?

Engineering works in a utility manhole

## HOT WORKS AND FIRE SAFETY

Due to the nature of the work being carried out or the equipment/substances being used, fires are a real threat on construction sites or during certain construction works.

Employers must ensure that the potential for fires has been considered as part of the risk assessment process and that appropriate fire-fighting measures are in place, assembly points have been established, communication mechanisms are established and the workforce are aware of the procedures in place.

Where specific hot works are being carried out (e.g. cutting, welding using naked flame tools), it may be necessary to introduce additional controls (e.g. a permit to work system) to minimise the risk of a fire starting or spreading.

Grinding operation creating sparks which could start a fire

## EXERCISE 2

### Self-assessment questions

1. What are the four colours used for safety signs?

   a) ..................................................................................

   b) ..................................................................................

   c) ..................................................................................

   d) ..................................................................................

2. What are the two key action levels under the Noise at Work Regulations?

   a) 85 and 95dB
   b) 80 and 85dB
   c) 75 and 85dB
   d) 80 and 90dB

**3.** Can exposure to one very loud noise cause deafness?

a) Yes
b) No

**4.** List four different health and safely control measures that can be applied to reduce risk?

a) ..................................................................................

b) ..................................................................................

c) ..................................................................................

d) ..................................................................................

**5.** Which muscles should be used when lifting a load?

a) Arm muscles
b) Leg muscles
c) Back muscles
d) Stomach muscles

**6.** Give two examples of what can happen when using vibrating tools or equipment.

a) ..................................................................................

b) ..................................................................................

**7.** List two things that must be assessed when using work equipment.

a) ..................................................................................

b) ..................................................................................

You will find the answers on pages 107–108.

# MODULE THREE

# Accidents and incidents

# OCCUPATIONAL ILL HEALTH IS STILL A BIG ISSUE FOR WORKERS INVOLVED IN CONSTRUCTION WORK

Welcome to module three. In this module you learn more about accidents and incidents, how they occur and what needs to be done when they do occur.

## WHAT IS AN ACCIDENT/INCIDENT?

These are unplanned/uncontrolled events which have led to, or could lead to, damage or injury to individuals, property, plant or any other loss to an organisation.

Accidents and incidents occur as a result of unsafe acts and unsafe conditions which are affected by:

### Organisational factors

These are factors relating to the organisation you work for; for example:

- the safety culture within the organisation;
- management commitment to safety

### Human factors

These are factors that may relate to you; for example, your attitude, the training you have had, any disabilities you may have, your health and/or fitness.

## Occupational factors

These are factors relating to the work you do and the environment you do it in; for example:

- If you are using shovels to dig holes, there will be a greater risk of muscular strain due to the manual handling involved.
- If you work in a kitchen, you are more at risk from knife injuries and burns from touching hot surfaces.
- High noise levels in a factory can damage your hearing.
- Poor ventilation within a laundry or kitchen can cause heat stress.
- Poor lighting within an office can cause headaches and eyestrain.
- Working in a dusty atmosphere such as a commercial bakery can cause breathing problems.

Once things start to go wrong (i.e. we have an unsafe act and/or unsafe condition), they act like dominos. A chain reaction takes place that cannot be stopped, resulting in accidents and injury.

Domino effect leading to incidents

Each year there are thousands of accidents reported to the enforcement authorities. Some of the most common categories include:

- manual handling disorders

- slips, trips and falls
- being struck by a moving object.

Unfortunately 95% of all accidents are caused by human error and are therefore preventable.

**Q**

Can you think of any accidents that have recently occurred in your workplace? What do you think really caused them?

Once an accident, dangerous occurrence or near miss has occurred, it is vital that it is reported for the following reasons:

## 1. Legislation

The Reporting of Injuries, Diseases and Dangerous Occurrences Regulations (RIDDOR) require employers to report the following to the local enforcing authority within specified time limits:

- fatal accidents
- fractures (not in the hand or foot)
- amputations
- loss of sight
- any injury resulting in immediate hospitalisation for more than 24 hrs
- specified dangerous occurrences such as a collapsing scaffold
- explosions
- specified workplace diseases that occur as a result of work activity.

The most common reason for employers to report an accident is when the injured person is off work for more than seven days as a result of the accident.

You as the employee need to notify your employer of all accidents that occur to you so that he can in turn fulfil his legal obligations.

## 2. Claim/Claim defence

In order for an employee to make a claim from their employer, there must be documented evidence of the incident having occurred. The accident book is often used to ensure this process takes place.

The employer also needs to know that the incident occurred so that he has the opportunity to do something about it. When reporting an accident, as much information as possible must be recorded.

### 3. Hazard identification

By documenting and being aware of the incident, the employer is better able to ascertain the cause of the incident and to implement corrective action to prevent it happening again (this may include carrying out a risk assessment or amending an existing risk assessment).

### 4. Monitor trends

Having records of all accidents/incidents that have occurred in a place of work over a period of time allows your employer to:

- analyse the types of incidents that have occurred;
- identify the staff involved and any training needs;
- ascertain the frequency of the incidents;
- ascertain the seriousness of the incidents;
- compare results with other similar businesses;
- compare results with previous years.

Research (carried out by Bird) has shown that for every accident resulting in a major injury there were approximately 10

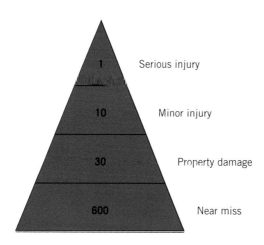

Bird's triangle

resulting in minor injuries, 30 with property damage and 600 near misses.

## WHY INVESTIGATE ACCIDENTS/INCIDENTS?

By investing time in investigating incidents your employer can:

- identify the cause of the accident;
- improve safety standards within the workplace;
- prevent any recurrence by implementing a corrective action plan;
- defend fraudulent claims (employees have up to three years to claim for compensation);
- minimise the risk of prosecution by external enforcing authorities.

## WHO SUFFERS FROM AN ACCIDENT?

The cost of having an accident can be counted in many ways. For example:

- loss of production
- plant/equipment damage
- physical harm
- psychological damage
- absenteeism
- loss of earnings
- permanent disability.

The real sufferers from the accident are the injured persons and their dependants. Always:

- **Think Safety** when carrying out your work
- **Work Safely** and you will be safe

• **Be Safe** and you will help avoid accidents involving yourself or others.

## HAZARDS IN THE WORKPLACE

Portable generator

Lifting operation using a crane

Worker on a ladder of insufficient length

Employee working at height

Poor storage of materials

Groundworks along a pavement

Worker in a confined space

A definition of a hazard is:

"SOMETHING (E.G. AN OBJECT, A PROPERTY OF A SUBSTANCE, A PHENOMENON OR AN ACTIVITY) THAT CAN CAUSE ADVERSE EFFECTS."

(**HSE** – 'Contains public sector information published by the Health and Safety Executive and licensed under the Open Government Licence v1.0'.)

In reality most things in the workplace or home have the potential to cause harm. For example:

- A single brick placed on a flat table in a room is relatively safe; however, the same brick falling from some high scaffolding could hit someone below and cause serious injury or even death.
- A strong acid in a properly labelled closed glass container in a laboratory cupboard would be relatively safe; however, the same acid left in a cup in a kitchen could be accidentally drunk, causing serious burns to the mouth and tracts.

A simple change in situation or circumstances can change the nature of the hazard from safe to very unsafe. This is known as the risk associated with the hazard.

The definition of risk is:

"THE LIKELIHOOD OF THE HAZARD CAUSING HARM AND THE PROBABLE OUTCOME."

(**HSE** – 'Contains public sector information published by the Health and Safety Executive and licensed under the Open Government Licence v1.0'.)

To help identification, hazards can be divided into five main categories within your workplace. These are:

1. **Physical:** These could be anything in the workplace you can feel or touch. Examples include:

   - doors
   - furniture
   - machinery
   - tools.

2. **Chemical:** These can be solid liquid or gas. Examples include:

   - detergents/soaps
   - acids/alkalis
   - correction fluid
   - bleach
   - glue
   - oils
   - paints
   - solvents.

3. **Biological:** This could be bacterial material on work surfaces. Examples include:

   - bacteria
   - moulds.

4. **Ergonomic:** Relates to the ill-health effects of poorly designed tasks and workstations leading to musculoskeletal problems including work-related upper limb disorders, whole body or hand/arm vibration syndrome.

5. **Psychological:** This includes mental health, workplace stress, violence at work, smoking and drug/alcohol abuse.

Hazards can be managed or controlled in a number of ways. The following list of control measures (in order of preference) can be targeted at any hazard.

- Ideally we should aim to eliminate hazards all together by removing the hazard or changing the process.
- Substituting hazardous materials or processes for less hazardous ones.
- Keeping people away from the hazard or the hazard away from the people (i.e. preventing exposure).
- Limiting time spent in the hazardous area or carrying out the hazardous task helps limit the exposure to a hazard.
- Introducing local exhaust ventilation can help remove the hazard if it is gaseous in nature
- Improving general ventilation in the workplace can help dilute the hazard if it is gaseous in nature.
- Improving housekeeping in the workplace can remove a number of tripping and falling hazards.
- Training employees to carry out the job correctly and safely; being fully aware of the hazards will reduce the likelihood of accidents and injury.
- Ensuring employees have adequate welfare facilities can help to reduce the hazards.
- Where necessary and available, carrying out health screening can help ensure other control measures are working and detect problems early.
- Providing employees personal protective equipment can help them reduce their exposure to the hazard providing that they are using the equipment in the right way.

**Q**

Could you, or does your employer, apply any of the above controls in your workplace?

**Q**

How would you describe your workplace? What types of hazard are you exposed to?

# FIRST AID

No matter how safe your place of work is there is always the potential for someone to have an accident or fall ill. This is why first aid forms a vital part of any employer's health and safety management system.

First-aid sign

The Health and Safety (First Aid) Regulations were intended to ensure that in the event of an incident, a qualified person would be able to preserve life and prevent the deterioration of someone who has fallen ill or is the victim of an accident.

Your employer must assess the risks within your workplace and provide adequate cover. In principle, your employer is required to have at least one fully qualified first aider per 50 employees (or emergency first aider as appropriate). That said, adequate cover must be provided during normal working hours taking into account shift patterns, holiday and sickness cover, the size of the location and hazards present. High-risk environments such as construction sites may require further specialist training for the first aider or even a dedicated first-aid room. In very low-risk environments, an appointed person may be sufficient. It would be their responsibility to provide basic medical cover, call for medical assistance and maintain any medical supplies on site.

It is important that you ensure you know who your first aiders are, where their normal place of work is and how to contact them. In the event of an emergency, it could help save lives. If you are not sure ask your supervisor/manager.

First-aid kits must contain as a minimum: individually wrapped plasters, sterile eye pads, triangular bandages,

sterile dressings and safety pins. First aiders are not authorised to issue or apply creams, pills, lotions, sprays or any
other medicines.

**Q**

Do you know who your local first aider is, where they are based and
how to contact them?

# EXERCISE 3

## Self-assessment questions

1. What is the definition of risk?

   ...........................................................................................

2. Accidents and incidents occur as a result of what?

   ...........................................................................................

   ...........................................................................................

3. Give two examples of accidents that would be reportable to the enforcing authority.

   a) ....................................................................................

   b) ....................................................................................

4. List three different reasons for investigating accidents.

   a) ....................................................................................

   b) ....................................................................................

   c) ....................................................................................

5. What is the definition of a hazard?

   ...........................................................................................

   ...........................................................................................

   ...........................................................................................

   ...........................................................................................

**6.** List four of the five hazard categories.

a) ........................................................................................

b) ........................................................................................

c) ........................................................................................

d) ........................................................................................

**7.** What are the two key purposes of first aid?

a) ........................................................................................

b) ........................................................................................

You will find the answers on pages 108–109.

# Proactive health and safety

# HSE CONFIRMS THAT THERE WERE NO WORK-RELATED FATALITIES DURING THE PREPARATION AND BUILDING OF ANY OF THE OLYMPIC VENUES FOR LONDON 2012

Welcome to module four. In this module you will learn about some of the practical steps you and your employer can take to help improve health and safety with construction activities and therefore reduce the risk of accidents and ill health. Remember your employer can only succeed with your help, co-operation and commitment.

## HEALTH AND SAFETY POLICY AND RESPONSIBILITIES

Your employer must produce a written policy statement detailing his/her commitment to ensuring yours and other people's health, safety and welfare. In addition he/she must set out clear responsibilities and accountabilities for key members of staff to ensure that safety is really put into practice and make arrangements to ensure the correct selection of contractors and that all safety procedures developed are implemented, monitored and reviewed as necessary.

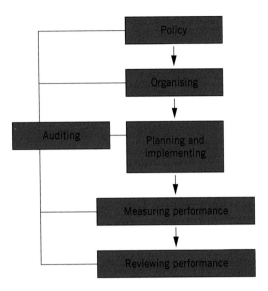

Successful health and safety management (HSG65)

**Q**

Have you seen and read your company's health and safety policy statement?

# PLANNING

Planning is one of the key elements in construction, proven to reduce the potential for accidents and incidents. Planning is necessary from the client's, designer's and contractor's perspective to ensure that all potential issues are identified, assessed for risk and managed prior to starting any work. Planning should take place pre-construction, during construction and post-construction.

# INSPECTIONS

Health and safety inspections may be carried out using checklists and/or proformas which measure conformance to the company's health and safety procedures in terms of documentation and the physical hazards present in the workplace. This allows your employer to "bench mark" and highlight the improvement or falling of standards ("performance tracking"). You may be asked to assist in the inspection process.

In reality you can carry out an inspection of your local place of work every day on an informal basis to satisfy yourself and by making your employer aware of any hazards you could help to prevent accidents.

Some equipment being used as part of the construction work may be subject to statutory inspections (e.g. scaffolding or lifting equipment).

# AUDITS

Your employer may from time to time carry out audits to measure the effectiveness of the procedures and processes he has in place. These can help highlight deficiencies and gaps within his overall construction safety management system. The audits may be carried out by internal staff or external companies.

## HAZARD/NEAR MISS REPORTING

As stated, your assistance is vital to improve safety standards. Always report any hazards, near misses or dangerous situations you see or are aware of. In some instances you may even be able to resolve the issues yourself and help prevent an accident. Your employer may have a formal hazard/near miss reporting system in place. If so use it as it could help save a life or prevent an accident.

## TRAINING

The purpose of training is to ensure that the employee is able to carry out the task/work activity to be performed consistently to an agreed standard and in a safe manner. All training, whatever the task, should cover any safety measures needed to ensure the task is performed safely (e.g. the correct use of PPE, manual handling techniques, machine operating).

Your employer may deliver the training on or off the job; it may be formal or informal, internal or external, in a group or individual basis. However, when training is conducted, the purpose is to ensure that you learn. Your employer should document details of any training provided to prove that it did take place.

Examples of when training is necessary include:

- induction training for all new employees to the business or new department/section
- refresher training
- specific training (e.g. first aid)
- training when new equipment is introduced.

Classroom-based training

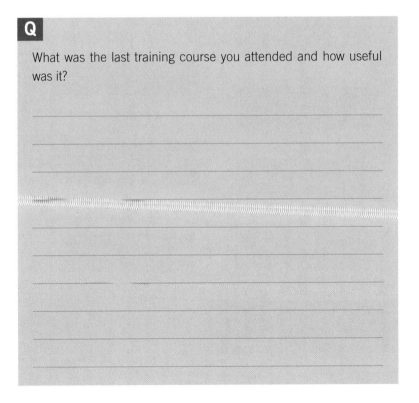

**Q**

What was the last training course you attended and how useful was it?

# SAFE SYSTEMS OF WORK/RISK ASSESSMENTS

A safe system of work is a formal written procedure which results from a systematic examination of a task or work activity in order to identify all the hazards and associated

**The Process of Risk Assessment**

The risk assessment process flowchart

risks involved in the task or activity (risk assessment). Once completed it should be used to provide you with instruction in the safe method of work and to ensure that the hazards are eliminated or the remaining risks are minimised.

The risk assessment identifies the key steps within the activity (how it is done), all the potential hazards are then identified (the "what if" chain of events) and assessed for risk. This is then followed by an action plan of any preventative measures needed to eliminate or reduce the risks (the "what can be done to prevent accidents" part of the assessment).

Your employer must complete risk assessments for all tasks carried out, both routine and abnormal. Always ensure that you have been made aware of the relevant risks associated with the work you do and that you are aware of any particular precautions that must be taken.

## PLANT/EQUIPMENT MAINTENANCE

Your employer has a duty to carry out or have carried out on their behalf specific tests and inspections to ensure that equipment/plant is safe to operate. Before using any plant or equipment, ensure you have the necessary training/knowledge and always carry out a quick visual inspection of the plant or equipment prior to use. Some equipment will be subject to statutory inspections (e.g. lifting equipment).

By developing and implementing a planned maintenance programme, your employer cannot only comply with legislation but also help ensure that work equipment is likely to be safe to use when properly used.

## COMMUNICATION

Your employer has a legal duty to consult with you on matters relating to health and safety directly or through nominated representatives (these may be unionised workplace representatives or non-unionised workplace representatives). It is in his interest and yours to ensure that you are fully aware of all hazards and risks present within the workplace.

Remember communication is a two-way process. It may be achieved through many ways such as:

- formal meetings
- informal meetings
- newsletters
- tool box talks
- training sessions
- notice boards
- emails
- posters.

Ensure you know who your representatives are and how you can communicate matters relating to health and safety to your employer.

Information board on a construction site

# HEALTH SURVEILLANCE/ OCCUPATIONAL HEALTH

Construction work has been going on for thousands of years and has been impacting workers' health and safety in terms of injuries and diseases occurring.

Although legacy illnesses such as deafness, asbestosis and lung cancer that take years to manifest themselves are reducing (although they are still present), a number of new hazards have appeared (e.g. MDF dusts, work-related stress, hand and arm vibration) and some are more or less constant (e.g. musculoskeletal disorders).

A worker's health will be directly affected by the type of environment he is working in. Occupational health surveillance is an important tool that can be used by an employer as part of his management strategy to help monitor control measures in place and identify and diagnose potential health problems within the workforce early on so that they can be treated.

Due to the nature of the work and the substances you may be exposed to, your employer may carry out some health surveillance to try to identify potential problems as quickly as possible and prevent them causing long-term illness. Examples include:

- audiometric tests for people working in noisy environments;
- blood lead levels for people working with lead;
- lung function tests for people working with asbestos or silica/wood dusts;
- hand examinations as a result of exposure to chemicals/vibration.

Unfortunately because occupational ill health issues take many years to manifest themselves, they are often ignored by employers and employees do not always realise they have a problem until it is too late.

It is also important to note that as construction work can be physically demanding, it is important that employees eat sensibly and on time to avoid poor diet-related issues (fatigue, diabetes, obesity).

## CO-OPERATION

The management of health and safety is a two-way process requiring the commitment and co-operation of both employer and employee.

You can help your employer improve safety standards and reduce accidents by the positive actions you take.

## EXERCISE 4

### Summary questions

To assess your level of understanding please complete the following exercise.

1. Give two examples of how inspections can help manage health and safety.

   a) .................................................................................

   b) .................................................................................

2. Is there a legal requirement to carry out risk assessments?

   a) Yes
   b) No

3. List three different ways of delivering training.

   a) ...........................................................................

   b) ...........................................................................

   c) ...........................................................................

4. List three ways the employer can communicate with employees.

   a) ...........................................................................

   b) ...........................................................................

   c) ...........................................................................

5. Why is it important to report hazards and near misses?

   ...........................................................................

6. Does your employer have to produce a safety policy statement?

   a) Yes
   b) No

7. When is it necessary to provide training?

   ...........................................................................

You will find the answers on pages 109–110.

# Answers to questions

# MODULE 1 – Construction health and safety law and enforcement

**1.** What is the maximum fine you can receive at a magistrates' court?

c) £20,000

**2.** What are the two important types of law?

c) Civil & Criminal

**3.** Can an enforcement officer take equipment away from your workplace?

a) Yes

**4.** List four reasons for employers to address construction health and safety.

Moral, Legal/Law, Financial/Cost, Business need

**5.** List two parties who have specific responsibilities under the Construction (Design and Management) Regulations.

Clients, designers, contractors

**6.** Who enforces construction health and safety law in the UK?

c) HSE inspectors

**7.** Health and safety law places a duty on:

c) Employers, employees and the self-employed

# MODULE 2 – Construction work issues

**1.** What are the four colours used for safety signs?

Blue, Yellow, Red and Green

**2.** What are the two key action levels under the Noise at Work Regulations?

b) 80 and 85dB

**3.** Can exposure to one very loud noise cause deafness?

a) Yes

**4.** List four different health and safety control measures that can be applied to reduce risk.

Elimination, substitution, limit exposure, local exhaust ventilation, health surveillance, general ventilation, housekeeping

**5.** Which muscles should be used when lifting a load?

b) Leg muscles

**6.** Give two examples of what can happen when using vibrating tools or equipment.

Vibration white finger, whole body vibration, damage to internal organs

**7.** List two things that must be assessed when using work equipment.

Answers could include the following; however, this list is not exhaustive: mechanical hazards (cutting, stabbing, drawing in, friction, entanglement, entrapment), non-mechanical hazards (noise, electricity, chemical, ergonomic, fire, radiation, explosion, lighting), stability, environment

# MODULE 3 – Accidents and incidents

**1.** What is the definition of risk?

The likelihood of a hazard causing harm and the probable consequences

**2.** Accidents and incidents occur as a result of what?

Unsafe acts, unsafe conditions

**3.** Give two examples of accidents that would be reportable to the enforcing authority.

Fractures, amputations, death, over three-day absence injury, 24-hour hospitalisation, unconsciousness, loss of sight

**4.** List three different reasons for investigating accidents.

Identify cause, prevent recurrence, claim defence, insurer stipulation

**5.** What is the definition of a hazard?

Something that can cause adverse effects

**6.** List four of the five hazard categories.

Physical, chemical, biological and ergonomic or psychological

**7.** What are the two key purposes of first aid?

Preserve life; prevent the deterioration of the casualty

# MODULE 4 – Proactive health and safety

**1.** Give two examples of how inspections can help manage health and safety.

Identify hazards, prevent accidents, demonstrate commitment

**2.** Is there a legal requirement to carry out risk assessments?

Yes

**3.** List three different ways of delivering training.

One to one, on the job, off the job, group training, distance learning, computer-based training, verbal instructions

**4.** List three ways the employer can communicate with employees.

Tool box talks, notice boards, newsletters, verbal instructions, meetings, videos

**5.** Why is it important to report hazards and near misses?

To prevent them from becoming accidents in the future

**6.** Does your employer have to produce a safety policy statement?

Yes

**7.** When is it necessary to provide training?

Induction, change of job, promotion, refresher, new equipment or change in process

# Glossary/Definitions

**Accident** An unplanned and/or uncontrolled event which has led to injury, damage or other loss to the business

**COSHH** Control of Substances Hazardous to Health Regulations

**dB** Decibel, the measure of sound

**DSE** Health and Safety (Display Screen Equipment) Regulations

**EHO** Environmental health officer

**EMAS** Employment Medical Advisory committee

**EPA** Environmental Protection Act

**FFI** Fee for intervention

**Hazard** Something (e.g. an object, a property of a substance, a phenomenon or an activity) that can cause adverse effects

**H&S** Health and safety

**HSC** Health and Safety Commission

**HSE** Health and Safety Executive

**HSWA** Health and Safety at Work Act

**MHOR** Manual Handling Operations Regulations

**MHSWR** Management of Health and Safety at Work Regulations

**MSDS** Material safety data sheet

**PAT** Portable appliance testing

**PPE** Personal protective equipment

**PUWER** Provision and Use of Work Equipment Regulations

**RIDDOR** Reporting of Injuries, Diseases and Dangerous Occurrences Regulations

**Risk** The probability and consequences of the hazard occurring

**RPE** Respiratory protective equipment

**SHE** Safety, health and environment

**Six pack** Term given to a group of six regulations introduced in 1993. They are the **MHSWR, MHOR, PPE, DSE, PUWER, WHSWR**

**UK** United Kingdom

**VDU** Visual display unit

**WEL** Workplace exposure limit

**WHSWR** The Workplace (Health, Safety and Welfare) Regulations

**WRULDs** Work-related upper limb disorders

# Image credits

**Figure 1** Injured man © RTFigures

**Figure 2** Winning trophy © bioraven

**Figure 3** Employers and employees must work together © Tsyhun

**Figure 4** Reasonably practicable © Subash Ludhra

**Figure 5** Failing to manage your health and safety can cost you your business © Subash Ludhra

**Figure 6** Obey the law and your duties © My Portfolio

**Figure 7** Magistrates' court © Subash Ludhra

**Figure 8a** An example of a prohibition sign © Subash Ludhra

**Figure 8b** An example of a safe condition sign © Subash Ludhra

**Figure 8c** An example of a warning sign © Subash Ludhra

**Figure 8d** An example of a mandatory requirement sign © Subash Ludhra

**Figure 9a** Poor wiring connection © Subash Ludhra

**Figure 9b** Danger electricity warning sign © northallertonman

**Figure 10a** Potential for cement burns to the unprotected hands © Alison Hancock

**Figure 10b** Asbestos cement sheets dumped on a road © Subash Ludhra

**Figure 10c** Ear plugs © Kellis

**Figure 11** Vibration white finger © Graham Dunn, Alamy

**Figure 12a** Examples of PPE used on construction sites

**Figure 12b** PPE for welders © Subash Ludhra

**Figure 13** Construction site compound © Subash Ludhra

**Figure 14** Ladder access on a scaffold © Imfoto

**Figure 15**  Mobile tower with outriggers © 3drenderings

**Figure 16a**  An example of a mobile elevated working platform (MEWP) © Hiper com

**Figure 16b**  MEWP © SergeBertasiusPhotography

**Figure 17**  Walkway created at a construction site

**Figure 18a**  Barriered excavation with some basic support © Serrato

**Figure 18b**  Excavations being mechanically dug © Bertold Werkmann

**Figure 19**  Engineering works in a utility manhole © Shutterstock

**Figure 20**  Grinding operation creating sparks which could start a fire © nattanan726

**Figure 21**  Domino effect leading to incidents © Ljupco Smokovski, Shutterstock

**Figure 22**  Bird's triangle

**Figure 23a**  Portable generator © canoniroff

**Figure 23b**  Lifting operation using a crane © Detchart Sukchit

**Figure 23c**  Worker on a ladder of insufficient length © Jeff Banke

**Figure 23d**  Employee working at height © Subash Ludhra

**Figure 23e**  Poor storage of materials © Smikeymikey1

**Figure 23f**  Groundworks along a pavement © J and S Photography

**Figure 23g**  Worker in a confined space © Soullo

**Figure 24**  First-aid sign © Subash Ludhra

**Figure 25**  Successful health and safety management (HSG65)

**Figure 26**  Classroom-based training © Subash Ludhra

**Figure 27**  The risk assessment process flowchart © Subash Ludhra

**Figure 28**  Information board on a construction site © Subash Ludhra

# Examination

Once you have completed the guide you can choose to complete the examination.

Complete the following exam to test your international health and safety workplace knowledge. All candidates will be informed whether they have passed or failed via email, and candidates achieving a pass mark of 75% or above will receive a certificate.

To find out details of where to send your completed exam, or to request an electronic version, visit www.routledge.com/9780415835459.

*You have thirty minutes – good luck.*

| Your FULL Name: (PRINT) | | Employee Reference: | Employer Name: |
|---|---|---|---|
| Date of Birth: | | Your Email Address: | |
| Date of Examination: | | | |

Please indicate by putting a tick in the appropriate box(es) your choice of correct answer(s)

1. What are the **TWO** types of law relating to health and safety?

   a) Criminal ☐
   b) Statute ☐
   c) Civil ☐
   d) Common ☐

2. Under the HSW Act, what are the **TWO** enforcement notices that are issued?

   a) Confiscation order ☐
   b) Prohibition ☐
   c) Renewal ☐
   d) Improvement ☐

3. What is the maximum fine that can be imposed for a health and safety offence in a **magistrates' court**?

a) £5,000 ❒
b) £10,000 ❒
c) £20,000 ❒
d) £40,000 ❒

4. What is meant by the term **reasonably practicable**?

a) The balance between risk and health ❒
b) The balance between risk and cost ❒
c) The balance between cost and health ❒
d) The balance between cost and prosecution ❒

5. Which one of the following **is not** a duty of an employer under the Health and Safety at Work Act?

a) Providing information, instruction and training for
   employees ❒
b) Providing a safe place of work ❒
c) Providing safe systems of work ❒
d) Consulting with employees ❒

6. Which one of the following **is not** a duty of an employee under the Health and Safety at Work Act?

a) To buy personal protective equipment for themselves ❒
b) Not to interfere or misuse anything provided for health
   and safety ❒
c) Care for the health and safety of themselves and other
   people ❒
d) Co-operate with their employer ❒

7. Who **enforces** construction health and safety legislation in the UK?

a) HSE inspectors and firemen ☐
b) Policemen and firemen ☐
c) HSE inspectors ☐
d) Environmental health officers and firemen ☐

8. Who has specific responsibilities under the Construction (Design and Management) Regulations?

a) Principal contractors ☐
b) Clients ☐
c) Designers ☐
d) All of the above ☐

9. What **message** does a Red safety sign give?

a) Warning ☐
b) Mandatory ☐
c) Prohibition ☐
d) Safe condition ☐

10. What are the **two key** action levels under the Noise at Work Regulations?

a) 80dB and 85dB ☐
b) 85dB and 90dB ☐
c) 80dB and 90dB ☐
d) 75dB and 80dB ☐

11. Which **muscles** should be used when lifting a load?

   a) The arm muscles ☐
   b) The leg muscles ☐
   c) The back muscles ☐
   d) The stomach muscles ☐

12. What are the **main purposes** of first aid?

   a) To cure the casualties' illnesses ☐
   b) To preserve life and prevent the deterioration of the
      casualty ☐
   c) To injure the casualty ☐
   d) To perform miracles ☐

13. Which **four factors** must be taken into account under the
    manual handling operations regulations?

   a) The task, load, working environment and individual
      capability ☐
   b) The weight, size, temperature and dimensions ☐
   c) The temperature, room space, individual capability
      and colour ☐
   d) The task, load, shape and time ☐

14. What are the **dangers** associated with electricity?

   a) Death ☐
   b) Electrical burns and shock ☐
   c) Fires ☐
   d) All of the above ☐

15. What is the **most effective** way of controlling a hazard and its associated risk?

a) Prevent exposure to it ☐

b) Substitute it ☐

c) Eliminate it ☐

d) Provide personal protective equipment ☐

16. What can prolonged use of vibrating tools or equipment cause?

a) Vibration white finger ☐

b) Whole body vibration ☐

c) Specific organ damage ☐

d) All of the above ☐

17. Which **one** of the following injuries is **not** reportable to the enforcing authorities?

a) Death ☐

b) Fractured arm ☐

c) Amputated leg ☐

d) Fractured finger ☐

18. Employers can communicate with their employees by?

a) Giving tool box talks ☐

b) Showing videos ☐

c) Having information on notice boards ☐

d) All of the above ☐

19. What is the **definition** of **risk**?

    a) Something that will cause harm ❐
    b) Something that has the potential to cause harm ❐
    c) The probability of harm being caused and the likely
       consequences ❐
    d) Being struck by a falling brick ❐

20. What is the **definition** of a **hazard**?

    a) Something that will cause harm ❐
    b) Something that can cause adverse effects ❐
    c) The probability of harm being caused and the likely
       consequences ❐
    d) Being struck by a falling brick ❐

21. Which **one** of the following is **not** a typical example of occupational health monitoring?

    a) Audiometric testing ❐
    b) Lung function testing ❐
    c) Blood lead level testing ❐
    d) Employee weight checks ❐

22. Which one of the following is **not** a primary reason for investigating accidents?

    a) Defending a claim for compensation ❐
    b) Identifying the cause of the accident ❐
    c) Helping to prevent a reoccurrence of the accident ❐
    b) Apportioning blame on employees ❐

23. <u>How long</u> does an employee have to make a compensation claim following an accident at work?

   a) One year ☐
   b) Three years ☐
   c) Five years ☐
   d) Ten years ☐

24. What **three** elements are required to produce a fire?

   a) Heat, light and smoke ☐
   b) Heat, fuel and oxygen ☐
   c) Fuel, nitrogen and heat ☐
   d) Oxygen, light and air ☐

25. When planning for work at height, what do employers need to consider?

   a) How can we avoid working at height? ☐
   b) How can we prevent falls of persons and equipment? ☐
   c) How can we protect workers if they do fall? ☐
   d) How cheaply can we do the job? ☐

26. Which one of the following is **not** a proactive safety measure?

   a) Carrying out regular audits ☐
   b) Carrying out regular training ☐
   c) Carrying out regular workplace inspections ☐
   d) Investigating an accident ☐

27. When is it **not** appropriate for an employer to provide health and safety training for his staff?

   a) When starting a new job ❐
   b) After an accident ❐
   c) When a new machine is introduced to the workplace ❐
   d) When leaving the company ❐

28. Which of the following safety measures is an example of **reactive monitoring**?

   a) Carrying out regular audits ❐
   b) Carrying out regular training ❐
   c) Carrying out regular workplace inspections ❐
   d) The analysis of accident data ❐

29. According to Bird's triangle **how many** near misses equate to one serious injury?

   a) 500 ❐
   b) 600 ❐
   c) 450 ❐
   d) 700 ❐

30. When faced with a confined-space environment what should employers ensure?

   a) A safe system of work is in place ❐
   b) A risk assessment has been completed ❐
   c) Work is carried out quickly ❐
   d) Only one person enters the confined space ❐

Well done, you have now finished the test. Please check to ensure that you have answered all 30 questions (remember some questions require two answers).